peak plus start

Alan Brighouse/David Godber/Peter Patilla

Teacher's book

National Curriculum Guide:
Using and applying mathematics
(Mainly Levels 2 and 3)

Nelson

[redacted]

Thomas Nelson and Sons Ltd
Nelson House Mayfield Road
Walton-on-Thames Surrey
KT12 5PL UK

51 York Place
Edinburgh
EH1 3JD UK

Thomas Nelson (Hong Kong) Ltd
Toppan Building 10/F
22A Westlands Road
Quarry Bay Hong Kong

Thomas Nelson Australia
102 Dodds Street
South Melbourne
Victoria 3205

Nelson Canada
1120 Birchmount Road
Scarborough Ontario
M1K 5G4 Canada

First published by Thomas Nelson & Sons Ltd 1988

ISBN 0-17-421515-0

NPN 9 8 7 6 5 4

Filmset in the Nelson Teaching Alphabet by
Mould Type Foundry Ltd
Dunkirk Lane Leyland Preston England

Printed and bound in Hong Kong

Series design Richard Crawford Graphic Design

General Introduction

The Peak Plus series of pupils' books (each with an associated Teacher's Handbook) contains activities ranging from Level 2 to Level 6 of the National Curriculum. The series focuses on Attainment Targets 1 and 9 by giving children the opportunity to use and apply Number, Algebra, Measures, Shape and Space, and Data Handling in a practical way. The varied activities include investigations of mathematical situations, solving of problems, mathematical puzzles and games.

What is the aim of the series?
The aim behind this series is to provide for you, both through pupil materials and handbook support, activities which will give children the opportunity to work within Attainment Targets 1 and 9 in order to meet the following statements of attainment:

Level 2 – select the materials and the mathematics to use for a task;
– describe current work, record findings and check results;
– ask and respond to the question: 'What would happen if ...' ?

Level 3 – select the materials and mathematics to use for a task, check results and consider whether they are sensible;
– explain work that has been done and record findings systematically;
– make and test predictions.

Level 4 – select the materials and the mathematics to use for a task, plan work methodically;
– record findings and present them in oral, written and visual form as appropriate;
– use examples to test statements or definitions.

Level 5 – select the materials and the mathematics to use for a task, check there is sufficient information, work methodically and review progress;
– interpret mathematical information presented in oral, written or visual form;
– make and test simple statements.

Level 6 – design a task and select appropriate mathematics and resources, check there is sufficient information and obtain any that is missing; use trial and improvement methods;
– use oral, written and visual forms to record and present findings;
– make and test a generalisation and simple hypothesis; define and reason in simple contexts with some precision.

Because of the investigative nature of the activities within the Peak Plus series, it is difficult to give an exact indication of the level of work within each book. However, as a guide, the following table may be helpful:

Peak Plus Starters Peak Plus More Starters	mainly levels 2 and 3
Peak Plus 1 Peak Plus 2	mainly levels 3 and 4
Peak Plus 3 Peak Plus 4	mainly levels 4, 5 and 6

The activities within the series are predominantly concerned with work where the answer is of less importance than the method of working, and where the real purpose is to give children the confidence to think for themselves. Undoubtedly, different children will gain from the activities in different ways: some will extend their investigative ability more than others, but we hope that they will all become more able to apply their skills to new mathematical situations in a systematic and thoughtful way. Out of the work we hope will develop an attitude of questioning and predicting, and a realisation that careful consideration of approach and methods usually resolves problems more easily and quickly.

The value of the series for children may be summarised as:

1. heightened enjoyment of mathematical activity;
2. discussion of problems with others (children/teachers);
3. increase in confidence through a sense of achievement;
4. encouragement of children to question and predict;
5. useful application of skills acquired (both recently and in the earlier years – in this way the work also has a revisionary value);
6. encouragement of a systematic approach to problems;
7. realisation that mathematics is concerned with procedure and methods as well as answers;
8. opportunity for children to work on a common problem (albeit at different levels);
9. encouragement to record appropriately as an important part of all mathematical activity.

What is the teacher's role in the use of this series?
The main aim of this series is to encourage the child to think mathematically, making your role much more of a 'sounding board', prompting where it is judicious and sensible so to do, but, in the main, allowing children to follow their own lines of exploration, and discussing with them the conclusions they draw. Obviously, there are areas where all children will need help. Some direction may well be needed in methods of recording, so that appropriate routines and good habits are established, but a far greater degree of responsibility than usual needs to be thrown on to the child in formulating decisions and following them through. There is always a temptation in this work to step in and tell children what to do – this is a temptation which, in the main, should be resisted, if children are to be encouraged to develop their own strategies and predictions. You are more an observer and prompter than teacher in this work, but you are also a valuable resource for children to draw upon when needed.

One responsibility, however, which does fall principally on you is the provision of adequate resources for the work. The enthusiasm of any child for investigative work falls flat if the necessary apparatus is not available, and easily accessible. To help with this provision, the apparatus needed is listed in each of the pupils' books.

How does the teacher organise the work?
If you are using the series to introduce, or increase the amount of, work of an investigative nature in the classroom, we would suggest the following as helpful advice:

1. Make the work fairly simple to start with. Do not worry if the work is based on skills long since acquired, since this is quite appropriate. It can still be very stretching for even the most able children meeting this approach for the first time.

2. Have the children work in pairs of small groups. One of the most important aspects of

problem-solving and investigative work is the discussion between children jointly involved in the activity. To deprive them of discussion with each other during the work is particularly damaging, even though you may spend time with them at various stages of the work. If investigations are being introduced for the first time, do not involve the whole class initially, but set one or two small groups working at a time, until the children become more accustomed to the work.

3. Discuss with the children the most appropriate method of recording and establish definite routines which you will expect throughout the work. Prompt children to record 'how' and 'why', rather than just record answers to the problems. Some recording (possibly in note form) will need to be done during investigations, as well as the final and more systematic recording done at the end of an activity.

4. Ensure that you have sufficient apparatus for all eventualities!

5. Encourage children to predict (and record their predictions) before testing. This encourages children to think carefully about the possibilities which exist in any mathematical situation.

6. Where children have explored avenues which have been 'dead-ends', avoid giving children the impression that they have wasted their time. Discuss with them the reasons for their results in order that they may explore more productive ideas, rather than simply pointing them in the right direction.

7. It is not necessary for an investigation or puzzle to be completed in one lesson. One of the advantages of this type of work is that it can be picked up at odd moments during a day as well as during a maths lesson.

8. Once a puzzle or investigation has been completed by a group of children, it can be presented to other children to look at, when they have also completed their work. It is often interesting for children to look at the way others have attacked the problem and compare it with their own methods.

9. Once investigations are an established part of mathematics for the children, it may be possible to allow them to select their own choice of activity from the books as an added incentive to the work.

How is the work assessed?
The work done by the children from this series of books cannot be assessed in the way in which most other mathematics work is assessed. Perhaps in the final analysis the level of a child's enthusiasm for, and expertise in, problem-solving is the best measure, but the following are suggested as 'pointers':

1. the quality of a child's discussion about the problem;
2. the ability to recognise the strategies needed to solve a problem systematically;
3. the ability to perceive patterns and to extend and test them;
4. the growth of confidence in tackling problems – that is the ability to think mathematically.

How are the Teacher's Handbooks to be used?
The purpose of the Teacher's Handbooks is to give you sufficient help and advice to develop investigatory work as an integral part of mathematics. Many of the pupil's activities do not place great mathematical demands on the pupil or yourself, except in exploiting all the possibilities

which exist and ensuring that the approach to the work is suitable. We hope that, with the information given in the Handbooks, you will feel that the value of the activities can be maximised, and that the security they offer will enable you and your pupils to enjoy the activities to the full.

Pupil recording

Investigations are often completed over several lessons; therefore it is sensible for children to keep a separate file for their work, which can easily be added to. Rough working can be kept together prior to it being copied up. Ring-binders, flat files with treasury tags, each having envelopes glued inside the covers, to hold the odds and ends of paper templates etc., are all suitable. A simple, but effective, home-made file can be made from thin card, as follows:

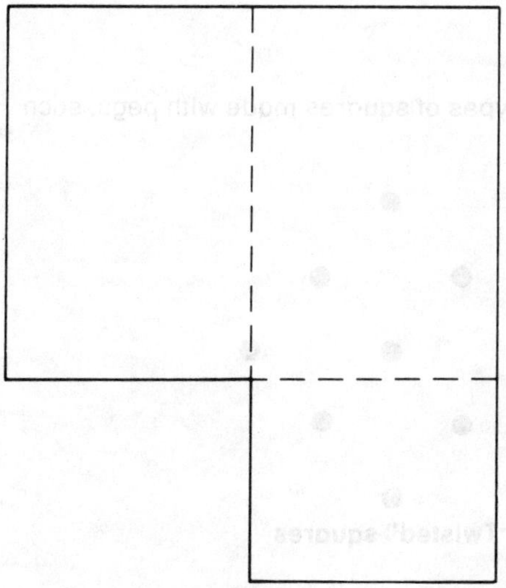

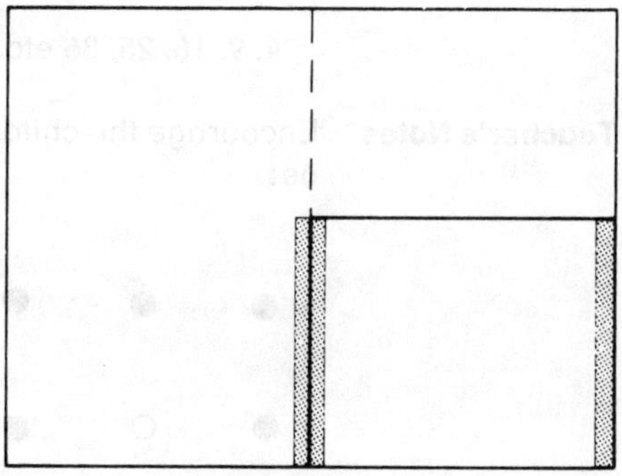

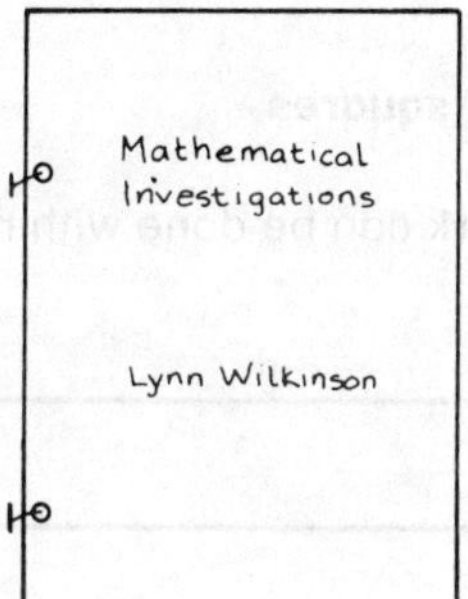

Peg Squares — Starters, Page 4

Activity Exploring squares on pegboards.

Apparatus Pegboard, pegs.

Knowledge Needed Recognition of squares.

Solutions The pattern of numbers of pegs needed for squares of increasing size is the pattern of square numbers, which is

4, 9, 16, 25, 36 etc.

Teacher's Notes Encourage the children to explore other types of squares made with pegs, such as:

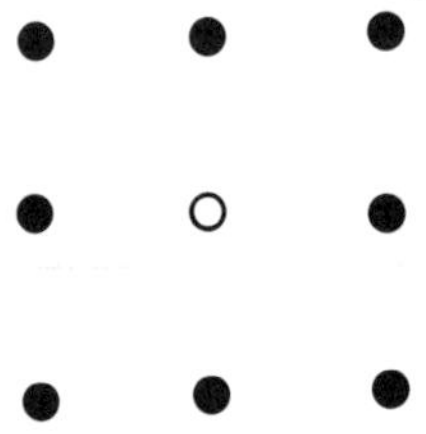

"Hollow" squares

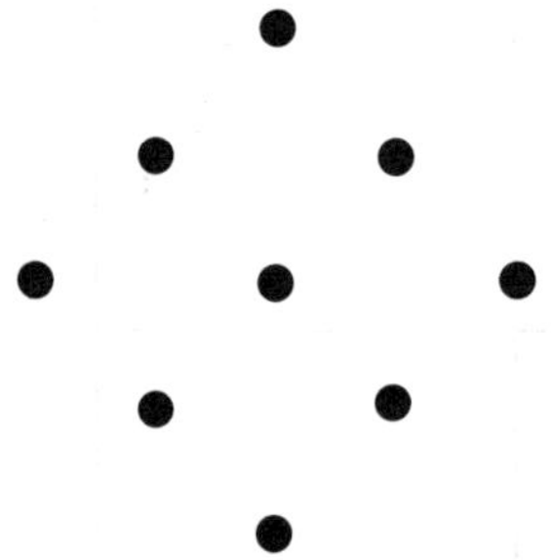

"Twisted" squares

Similar work can be done with rectangles and triangles of increasing size.

Matching Trains — Starters, Page 5

Activity Exploring sets of equivalent rods; early ideas of common factors, repeated addition, commutativity.

Apparatus Number rods (or number strips – see below).

Knowledge Needed Familiarity with number rods.

Solutions Three 5-rods match five 3-rods.
Four 7-rods match seven 4-rods.
Five 6-rods match six 5-rods.
Three 8-rods match eight 3-rods.

Teacher's Notes The rods shown in the pupil text are Cuisenaire rods, but Colourfactor can be used just as satisfactorily. If neither are available, card strips can be made which match the size of the rods.

Encourage the children to make other matching trains, choosing any two different-sized rods at random. When rods have a common factor (e.g. 4-rods and 6-rods) the matching is more frequent. When one rod is a factor of the other (e.g. 5-rods and 10-rods) the matching is at every multiple of the larger rod.

Extend the work by starting with a fixed train (e.g. four 3-rods) and find all other trains which will match.

Four 3-rods will match with:
Two 6-rods.
Three 4-rods.
Six 2-rods.
Twelve 1-rods.

Make it Balance — Starters, Pages 6–7

Activity Exploration of number relationships.

Apparatus Equaliser balance, washers.

Knowledge Needed Odd, even.

Solutions There are many solutions for each example.
The washers the children place on the balance can be put on either arm.

Teacher's Notes Discuss with the children the relationship between different combinations of odd and even, and the resultant total

e.g. Odd + Odd = Even
Odd + Even = Odd

Bead Lines — Starters, Page 8

Activities Finding different arrangements of coloured beads.

Apparatus Coloured beads and laces.

Solutions With three beads, there are six different arrangements.

Red – Green – Blue
Red – Blue – Green
Green – Red – Blue
Green – Blue – Red
Blue – Green – Red
Blue – Red – Green

Teacher's Notes Discussion is required with the children about appropriate methods of recording. These may include:

i. Keeping each lace of beads;
ii. Drawing the arrangement with coloured pencils;
iii. Recording the sequence of colours by name.

The work can be extended to include four beads.

Silver Coins

Starters, Page 9

Activity Combinations of silver coins.

Apparatus Silver coins (or plastic imitation coins).

Knowledge Needed Ability to total coins.

Solutions

For the ball costing 42p
Possible solutions include:

20p + 20p + 10p   20p + 10p + 10p + 5p

50p   10p + 10p + 10p + 10p + 10p

20p + 10p + 5p + 5p + 5p

For the book costing 99p
Possible solutions include:

50p + 50p   20p + 20p + 20p + 20p + 20p

50p + 20p + 20p + 10p   50p + 20p + 10p + 10p + 10p

For the toy costing 78p
Possible solutions include:

50p + 20p + 10p   50p + 20p + 20p

50p + 50p   50p + 20p + 5p + 5p

Teacher's Notes The main point of the activity is for the children to realise that some silver sets will not be appropriate when paying for particular items. For instance, when paying for the ball costing 42p, it is appropriate to offer a set of coins comprising 20p + 20p + 10p, but inappropriate to offer 20p + 20p + 5p + 5p. The same applies to the other items included in the pupils' text.

This work can be extended by asking for payment to be made with the fewest/most coins possible.

Activity Arranging number rods within an outline.

Apparatus Number rods (or number strips, see page 5).

Knowledge Needed Familiarity with number rods.

Solutions Accept any solutions which cover each shape exactly, e.g.

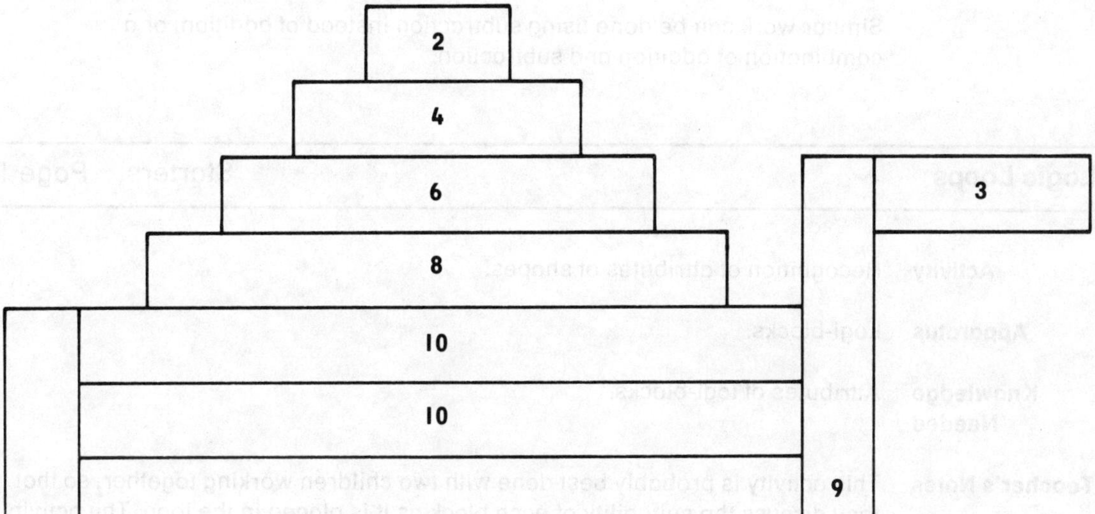

Teacher's Notes When covering the shapes, more than one rod of the same size can be used. For each shape there are many solutions; encourage the children to find those using fewest rods.

Children should also be encouraged to design and draw their own shapes for others to cover with rods.

Choose your Cards
Starters, Page 12

Activity Totalling with numeral cards.

Apparatus Numeral cards 1 to 10.

Knowledge Needed Familiarity with number bonds.

Teacher's Notes With each set of four numeral cards, encourage the children to find as many different answers as possible, using two, three or all four cards. Some choices of numeral cards will give fewer different answers than others.

Similar work can be done using subtraction instead of addition, or a combination of addition and subtraction.

Logic Loops
Starters, Page 13

Activity Recognition of attributes of shapes.

Apparatus Logi-blocks.

Knowledge Needed Attributes of logi-blocks.

Teacher's Notes This activity is probably best done with two children working together, so that they discuss the suitability of each block as it is placed in the loop. The activity can be done in two ways:

- **i.** Children select the logi-blocks as needed from a full set.
- **ii.** Children choose 8 logi-blocks initially, and try to arrange them. If this method is used, there may be occasions when it is not possible to complete the loop.

The work can be extended by asking children to make the loop with adjacent blocks which have two (or more) different attributes.

Spot Sets
Starters, Pages 14–15

Activity Totalling.

Apparatus Dominoes.

Knowledge Needed Ability to count.

Solutions **6-spot sets**
Accept any set where the spots total 6.

4-spot sets
Accept any set where the spots total 4.

e.g.

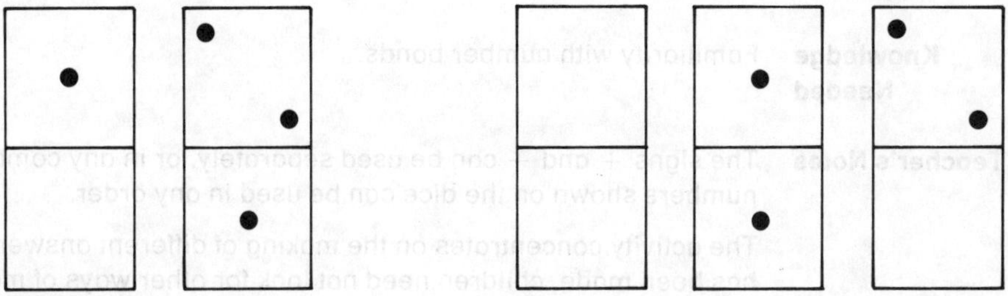

10-spot sets
The largest 10-spot set will have 5 dominoes.

12-spot set
The largest 12-spot set will have 6 dominoes.

Teacher's Notes Encourage the children to explore different ways of making a total with a set number of dominoes,

e.g. how many ways can you make a total of 12 spots with six dominoes?

Discussion will be needed on the best way to record the results.

Take it Away — Starters, Page 16

Activity Subtraction of numbers.

Apparatus Calculator.

Knowledge Needed Ability to use a calculator.

Teacher's Notes It will be necessary for the children to record each step as they go along.

With a starting display of 12, the fixed answer of 4 can be reached with a varying number of touches on the calculator.

e.g. [1] [2] [−] [8] [=] needs 5 touches

[1] [2] [−] [7] [−] [1] [=] needs 7 touches

Encourage the children to explore whether the fixed answer could be obtained with 6 touches, 8 touches etc.

The activity can be varied by starting with a fixed number on display and stating the number of touches allowed.
How many different answers can be obtained?

Dice Answers

Starters, Page 17

Activity Number operations.

Apparatus Dice.

Knowledge Needed Familiarity with number bonds.

Teacher's Notes The signs + and − can be used separately, or in any combination. The numbers shown on the dice can be used in any order.

The activity concentrates on the making of different answers; once an answer has been made, children need not look for other ways of making the same answer.

Coloured Flags

Starters, Pages 18–19

Activity Investigating arrangements of colours.

Apparatus Coloured pencils, squared paper, isometric paper.

Solutions **Colouring square flags, using red and blue pencils**
There are 16 different ways to colour the flag:

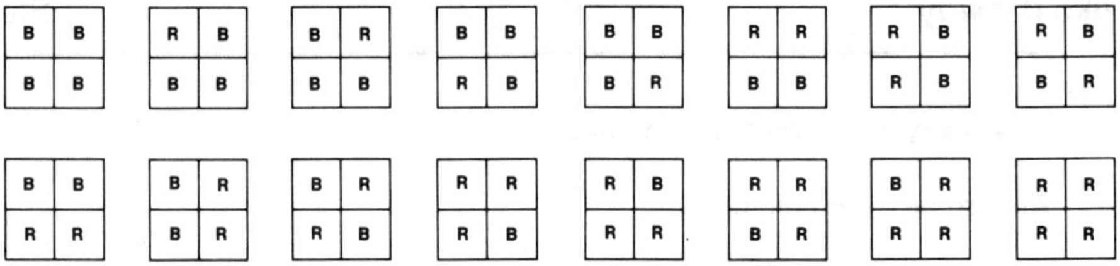

Colouring triangular flags, using green and yellow pencils
There are 16 different ways to colour the flag:

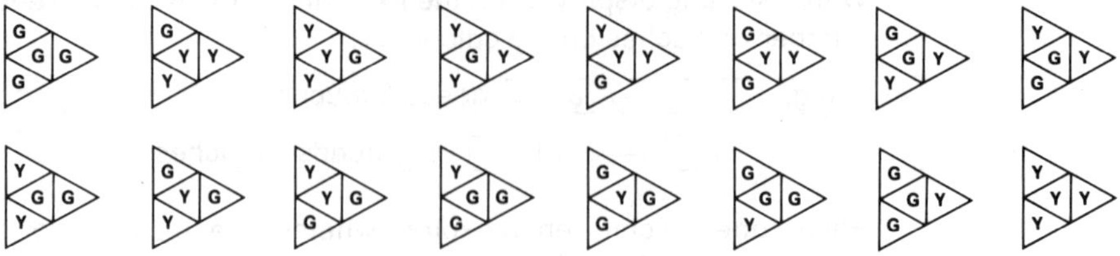

Teacher's Notes When children colour the flags, each individual segment (square or triangle) must be in one colour only.

Children may need prompting that every segment within the flag can be the same colour (i.e. the flag can be totally blue).

Encourage children to design their own flags and colour them in a similar way.

Three Coins

Starters, Page 20

Activity Investigating coin totals.

Apparatus Coins.

Knowledge Needed Values of coins. Ability to add coin values.

Solutions Amounts which cannot be made with three coins (up to 20p) are:
1p, 2p, 10p, 18p, 19p.

Teacher's Notes This activity can be varied by altering the number of coins and/or altering the upper limit.

Making Cubes

Starters, Page 21

Activity Exploring nets of a cube.

Apparatus Interlocking shapes (e.g. Clixi, Polydron).

Knowledge Needed Recognition of cube.

Solutions **Nets of a cube**

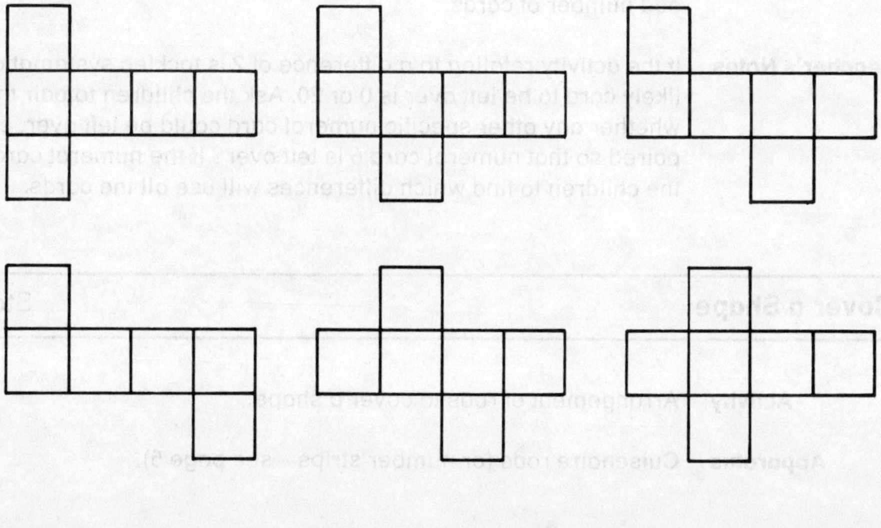

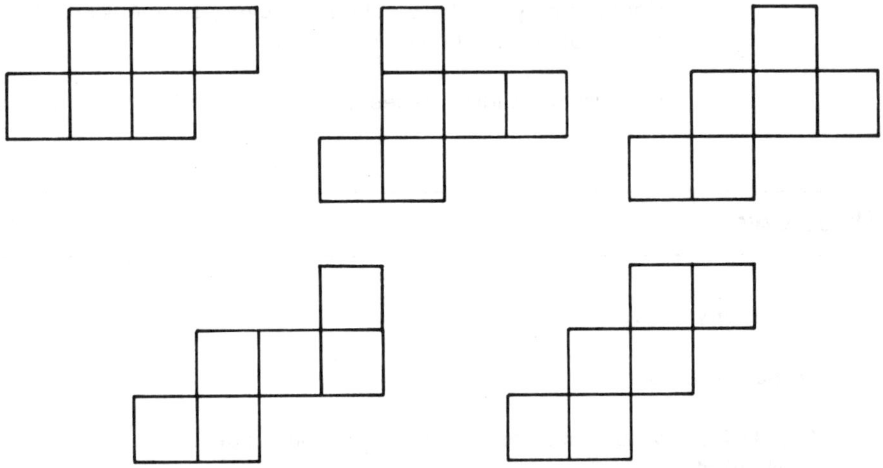

Teacher's Notes The nets of unfolded cubes can be copied on to squared paper. This activity can be reversed by the children starting with 6 interlocked squares and investigating whether they will fold to make a cube. Children may like to record those arrangements which cannot be folded into a cube.

Similar activities can be done with interlocking triangles.

What a Difference

Starters, Page 22

Activity Investigating number differences.

Apparatus Numeral cards 0 to 20.

Knowledge Needed Understanding of difference.

Solutions Using the number cards 0 to 20, at least one card will be left over, as there is an odd number of cards.

Teacher's Notes If the activity relating to a difference of 2 is tackled systematically, the most likely card to be left over is 0 or 20. Ask the children to pair the cards to see whether any other specific numeral card could be left over, e.g. can they be paired so that numeral card 6 is left over? If the numeral card 0 is omitted, ask the children to find which differences will use all the cards.

Cover a Shape

Starters, Page 23

Activity Arrangement of rods to cover a shape.

Apparatus Cuisenaire rods (or number strips – see page 5).

Knowledge Needed Familiarity with number rods.

Solutions There are several solutions, including:

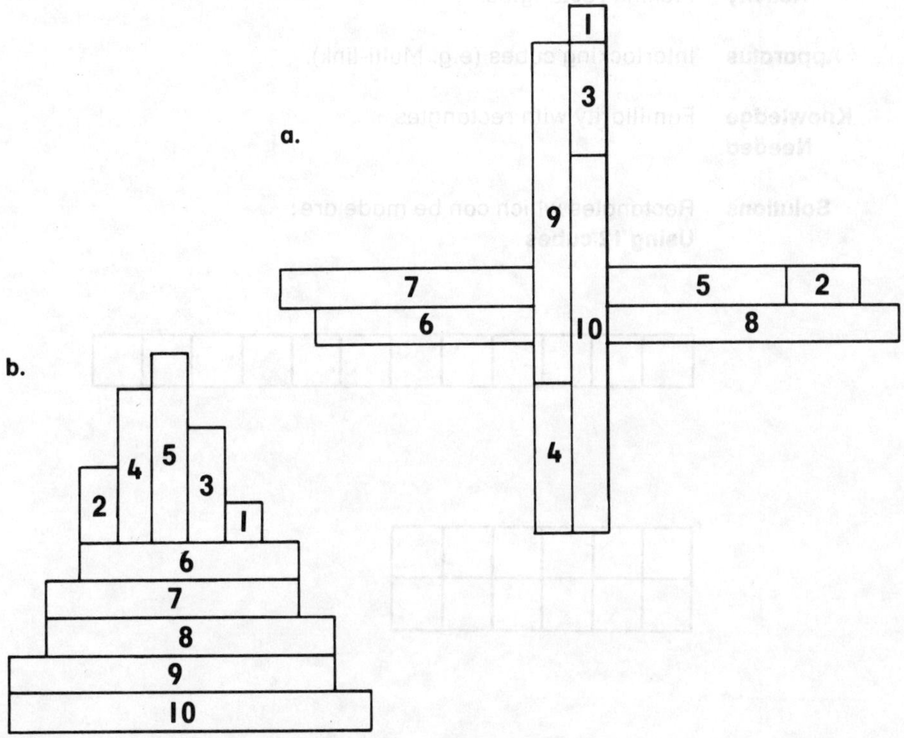

Teacher's Notes Children could copy the results of their investigations on to squared paper. Encourage them to find more than one solution for each shape. They could also create their own outlines, using rods 1 to 10, for other children to cover.

Elastic Shapes

Starters, Pages 24–25

Activity Exploring shape on a geo-board.

Apparatus Geo-board, elastic bands.

Teacher's Notes The main aim of this activity is to provide a basis for discussion about the similarities and differences of the shapes made on the geo-board. Discuss with the children the position and size of the shapes created, especially those which are identical but in different positions.

4-sided shapes
Encourage the children to make as many different 4-sided shapes as possible, and to name those which they can recognise (e.g. rectangle, square).

5-sided shapes
This activity allows children to meet pentagons other than the regular one.

Rectangles

Starters, Page 26

Activity Making rectangles.

Apparatus Interlocking cubes (e.g. Multi-link).

Knowledge Needed Familiarity with rectangles.

Solutions Rectangles which can be made are:

Using 12 cubes

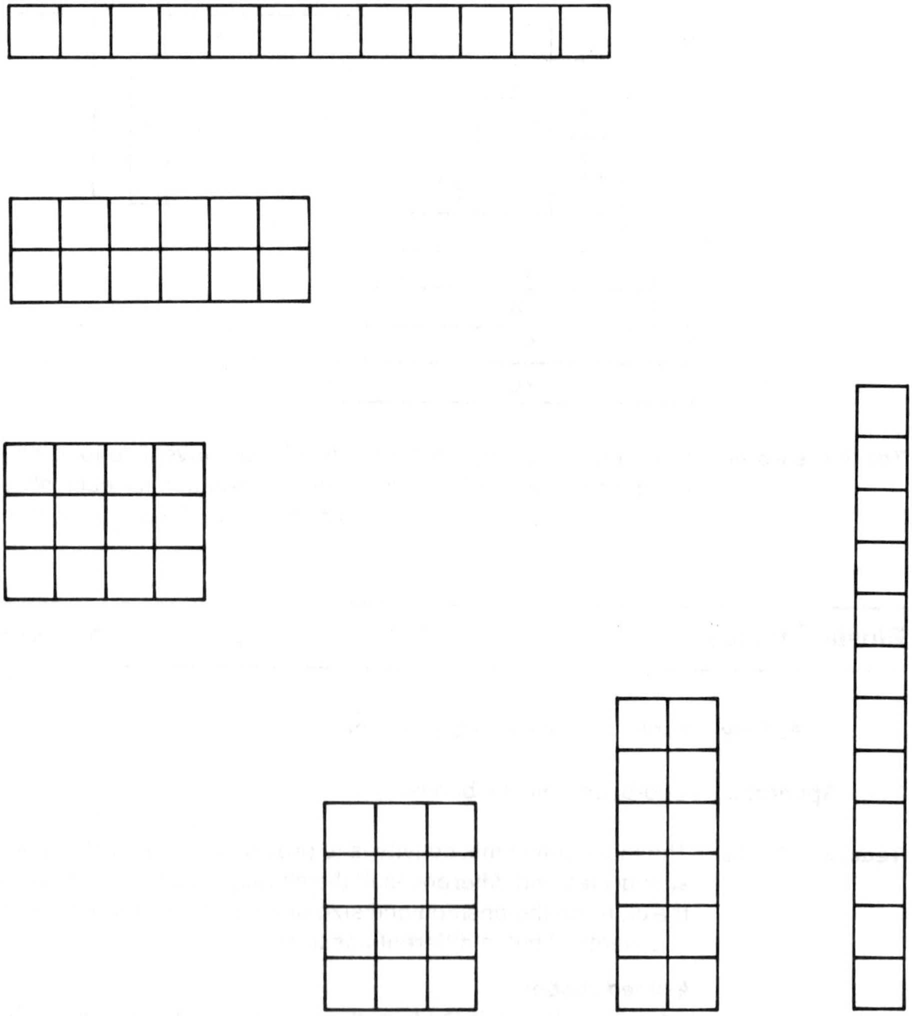

Using 16 cubes

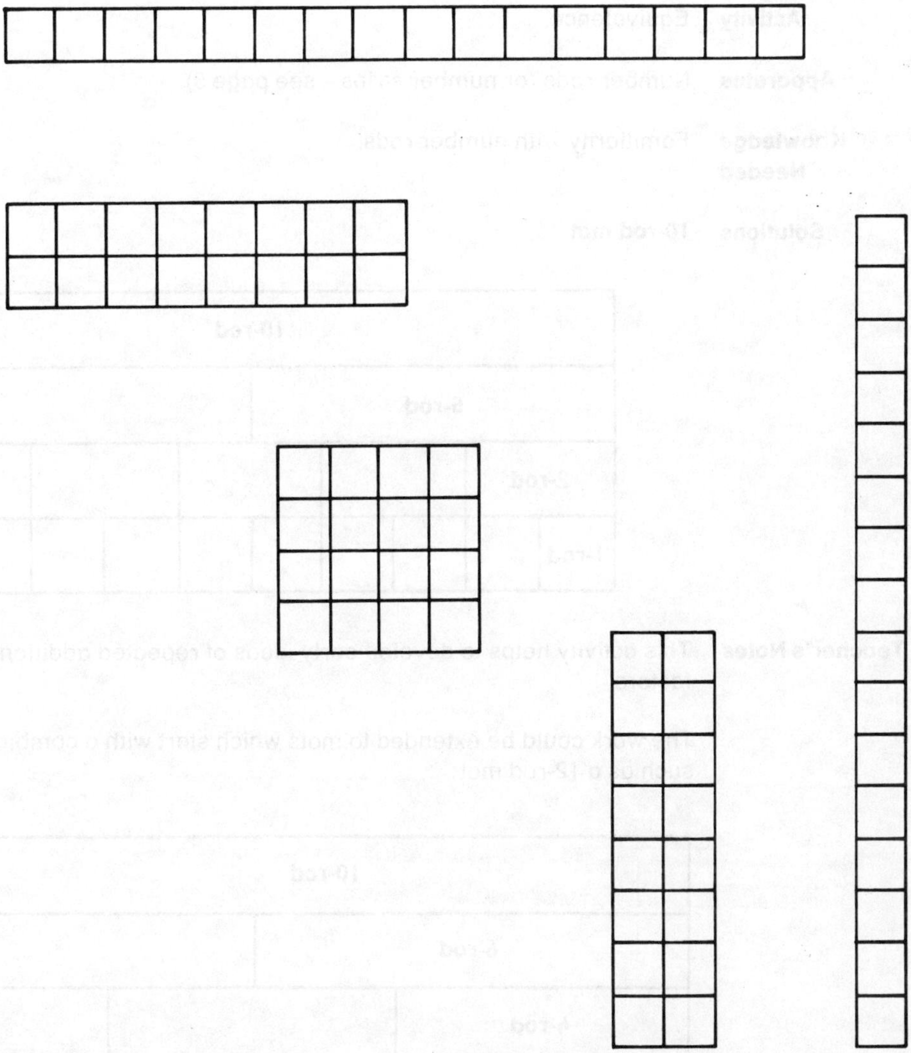

Teacher's Notes Some of the rectangles shown above are rotations; it will be necessary to decide whether rotations are to be accepted in the investigation. It may be necessary to ensure that children do not exclude the square as a solution – it can be accepted as a special rectangle.

Ask the children to try different numbers of cubes to form rectangles. They may be interested to investigate which numbers of cubes will make squares, and which will only make rectangles one unit wide (i.e. prime numbers).

Mats — Starters, Page 27

Activity Equivalence.

Apparatus Number rods (or number strips – see page 5).

Knowledge Needed Familiarity with number rods.

Solutions **10-rod mat**

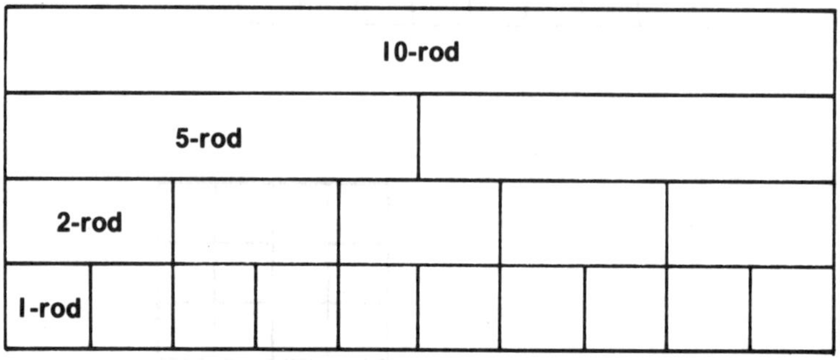

Teacher's Notes This activity helps to develop early ideas of repeated addition, multiples and factors.

The work could be extended to mats which start with a combination of rods, such as a 12-rod mat:

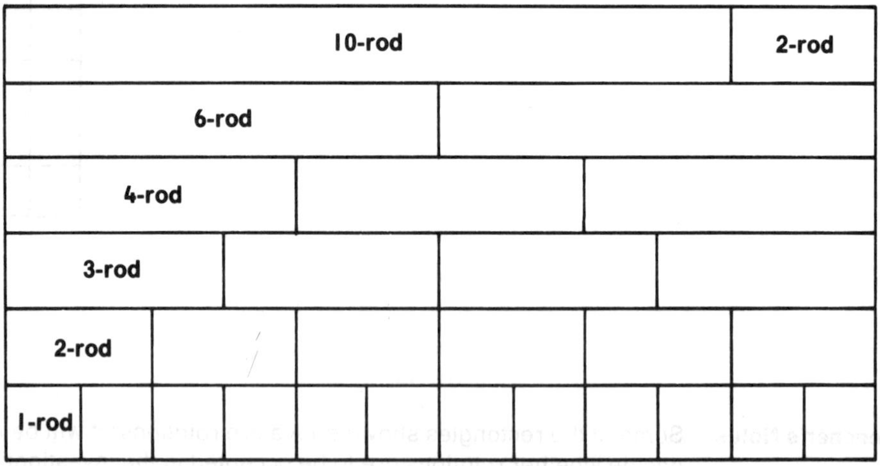

Dot Lines — Starters, Pages 28–29

Activity Investigation of dice totals.

Apparatus Dice.

Knowledge Needed Number bonds.

Solutions **With 3 dice**
The largest total you can make is 18.
The smallest total you can make is 3.
All totals from 3 to 18 can be made.

With 4 dice
The largest total you can make is 24.
The smallest total you can make is 4.
All totals from 4 to 24 can be made.

Teacher's Notes An alternative activity is to find different ways in which a specific total can be made. Other investigations could include:

Can the two visible lines of dots add up to odd and even numbers respectively?

What is the largest difference which can be created between the two visible lines of dots?

Addition Squares

Starters, Page 30

Activity Addition squares.

Knowledge Needed Number bonds.

Solutions **a.**

+	5	8
3	8	11
5	10	13

+	4	6
9	13	15
7	11	13

b. There are several solutions to the final two addition squares in the pupil's text. They include:

+	1	2
4	5	6
8	9	10

+	2	9
1	3	10
2	4	11

Teacher's Notes The final two addition squares in the pupil's text offer the opportunity to find as many different solutions as possible.

Making Boxes

Starters, Page 31

Activity Designing containers.

Apparatus Interlocking shapes (Clixi, Polydron etc), small balls, plastic cup, small tin, cone.

Teacher's Notes The containers made by the children can be either totally enclosed (i.e. with a lid), or open. Encourage the children to explore whether suitable containers can be made using only one shape, or whether they can be better made using a combination of shapes.

Cube Faces

Starters, Page 32

Activity Problem involving arrangements.

Apparatus Interlocking cubes (e.g. Multilink).

Teacher's Notes The activity can be extended by increasing the number of cubes and colours.

Domino Differences

Starters, Page 33

Activity Constant differences.

Apparatus Dominoes.

Knowledge Needed Understanding of difference.

Solutions **Difference of 3**

There are four dominoes with a difference of 3.
They are

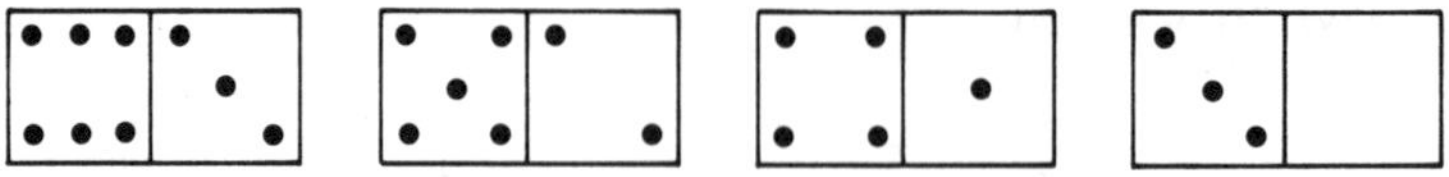

Difference of 2

There are five dominoes with a difference of 2.
They are

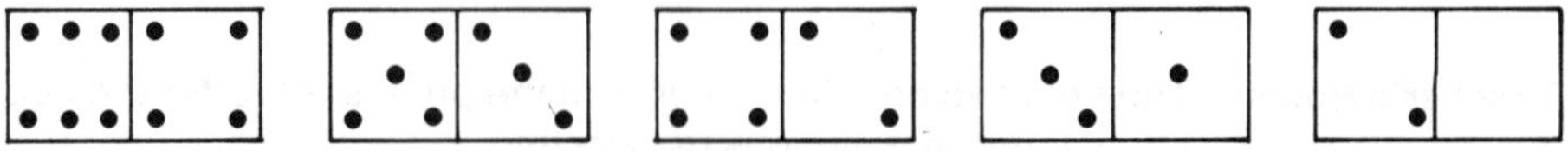

Teacher's Notes	Similar work can be done by looking at differences between whole dominoes, and finding pairs of dominoes with a stated difference.

Peg Patterns

Starters, Pages 34–35

Activity	Symmetry on a pegboard.
Apparatus	Pegboard, pegs.
Knowledge Needed	Ability to recognise symmetrical pattern.
Solutions	For each pattern there are many solutions. Check that the pattern is still symmetrical when the pegs have been taken away.
Teacher's Notes	Solutions could be copied on to spotty paper. If an odd number of pegs is taken away from a symmetrical pattern, at least one of them must be on the line of symmetry for the pattern to remain symmetrical.

Touch the Keys

Starters, Page 36

Activity	Calculator investigation.
Apparatus	Calculator.
Knowledge Needed	Ability to use a calculator. Number bonds.
Solutions	Check that each solution requires only five touches. e.g. [1] [2] [–] [4] [=]
Teacher's Notes	For further similar work, alter the number of touches and change the target number.

Rod Squares

Starters, Page 37

Activity	A number rod problem.
Apparatus	Cuisenaire rods (or number strips – see page 5).
Knowledge Needed	Familiarity with number rods.

Solutions The squares on the pupil's text can be covered with:

6-rods, 3-rods, 2-rods, 1-rods.

Teacher's Notes The square on the pupil's text is a 6 cm square. Give the children other sized squares to cover with rods in the same way.

On the Grid

Starters, Page 38

Activity Arrangement of dominoes on a grid.

Apparatus Dominoes.

Solutions A possible solution is:

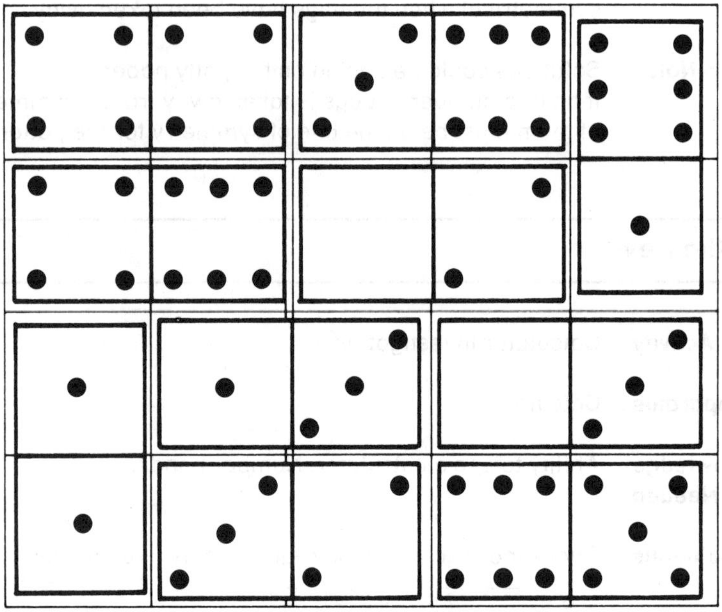

Teacher's Notes Children can design similar puzzles for others to solve; a collection of such puzzles can be made for the class.

Colour a Half

Starters, Page 39

Activity Investigating ways of halving.

Apparatus Coloured pencils, squared paper.

Knowlege Needed Half.

Solutions There are many different ways of colouring each of the shapes.

Teacher's Notes For each shape, encourage the children to find as many different ways as possible of colouring a half. Similar work can be done with shapes drawn on isometric paper.

Matching Totals

Starters, Page 40

Activity Totalling.

Apparatus Numeral cards 0 to 20.

Knowledge Needed Number bonds.

Solutions Accept any sets of numeral cards which total 15.

Teacher's Notes The number of cards in any set need not be restricted to three; any number of cards can be used providing their total is correct.

Silver Sets

Starters, Page 41

Activity Equivalence of coins.

Apparatus Silver coins (or plastic imitation ones).

Knowledge Needed Values of coins. Ability to total coins.

Solutions Accept any set which totals 50p.

Teacher's Notes Discuss with the children which sets have equal numbers of coins, which set has the largest/fewest number of coins etc. The activity can be repeated by changing the value of the equivalent sets to be made.

Coded Words

Starters, Pages 42–43

Activity Totalling.

Apparatus Calculator.

Knowledge Needed Ability to use a calculator.

Solutions Another "100 word" is WEDNESDAY. STRAWBERRY is worth 149.

Teacher's Notes Encourage the children to find the highest/lowest total for words within a theme (e.g. animals, days of week, names in class).

Also, ask the children to find the largest possible value for a three-letter word, and then extend to four-letter and five-letter words.

Odd Totals

Starters, Page 44

Activity Investigating odd and even totals.

Apparatus Numeral cards 0 to 10.

Knowledge Needed Number bonds, odd and even.

Teacher's Notes Children can classify their solutions into:

- 3-card solutions
- 4-card solutions
- 5-card solutions

Encourage the children to look at the combination of odd and even numbers within each set.

Spots on the Table

Starters, Page 45

Activity Dice problem.

Apparatus Dice.

Knowledge Needed Number bonds.

Solutions

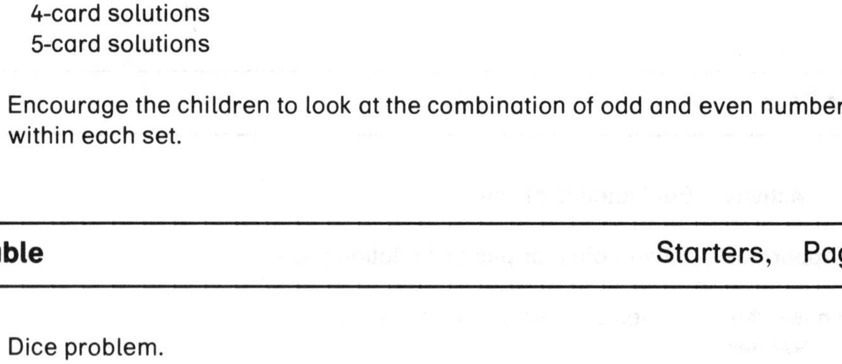

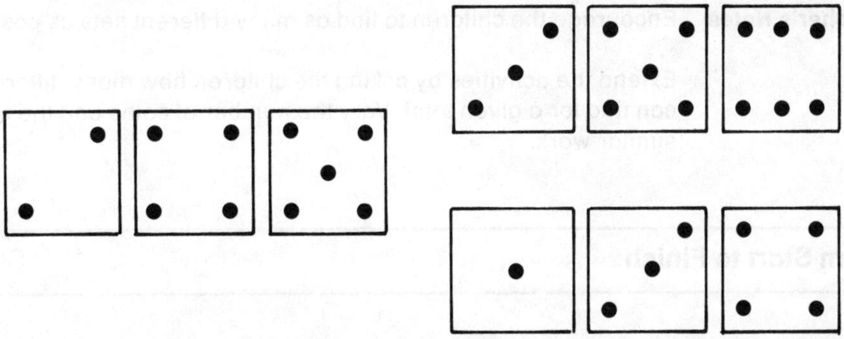

Teacher's Notes Children may need to model the pictures in the pupil's text, using dice, in order to find the solutions.

Some children may realise that opposite faces on dice total 7, and therefore the hidden numbers can be quickly worked out.

Money Sets

Starters, Page 46

Activity Equivalence of coins.

Apparatus Coins.

Knowledge Needed Value of coins.

Solutions **6p sets**

There are five different sets of coins which total 6p:

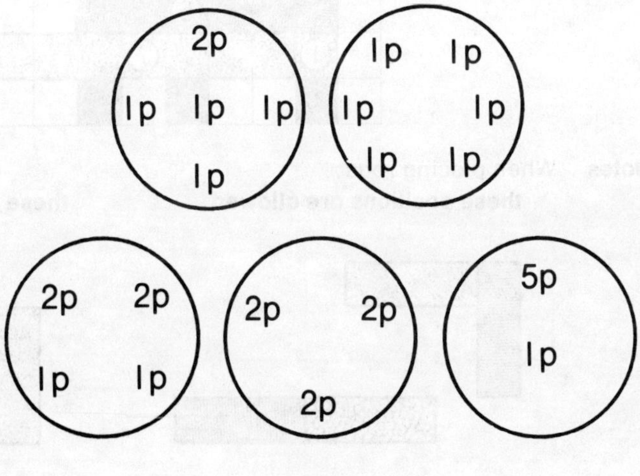

15p sets

Accept any sets of coins which total 15p.

Teacher's Notes Encourage the children to find as many different sets as possible for each value.

Extend the activities by asking the children how many different 5-coin sets they can find for a given total. Vary the number of coins and the given total for further similar work.

From Start to Finish — Starters, Page 47

Activity Arrangement of number rods on a grid.

Apparatus Cuisenaire rods (or number strips – see page 5).

Knowledge Needed Familiarity with number rods.

Solutions **A possible solution**

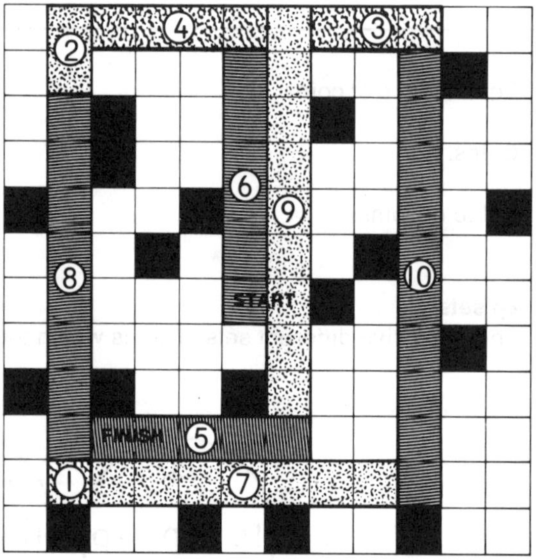

Teacher's Notes When placing rods:

these positions are allowed

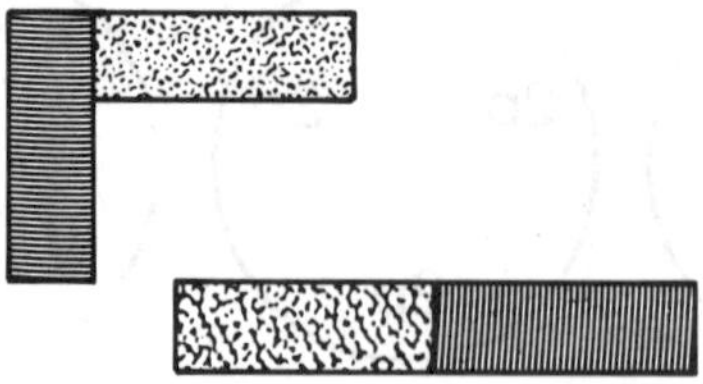

these positions are not allowed

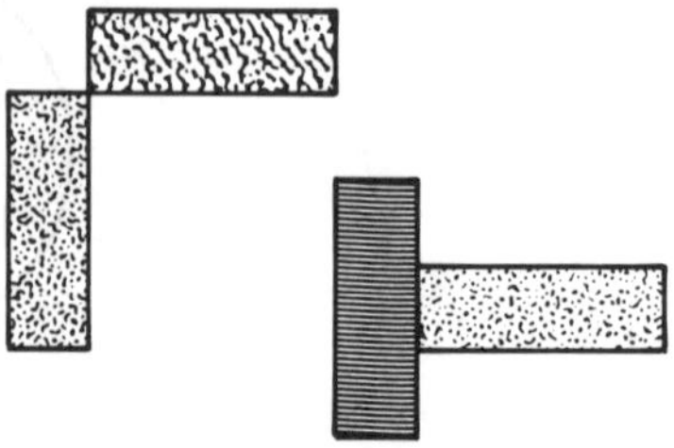

Encourage the children to produce their own routes to create a class collection.

Apparatus list

Balls (small)
Beads
Calculator
Clixi (interlocking shapes)
Coins
Coloured pencils
Cone
Cubes (interlocking)
Cuisenaire rods
Cup (plastic)
Dice
Dominoes
Elastic bands
Equaliser
Geo-board*
Lace
Logi-blocks
Numeral cards*
Pegboard*
Pegs
Tin (small)
Paper:
  Isometric paper*
  Squared paper*

* included in *New Peak Stationery Copymasters* by A. Brighouse, D. Godber and P. Patilla (Thomas Nelson, 1990)

Index

Arrangements 8, 10, 11, 18, 19, 23, 32, 38, 47
Calculator 16, 36, 42, 43
Codes 42, 43
Cubes 21, 32
Dice totals 28, 29, 45
Difference 22, 28, 33
Dominoes 14, 15, 33, 38
Equaliser 6, 7
Equivalence 5, 27, 37
Geo-board 24, 25
Halving 39
Logi-blocks 13
Logic 13
Money 9, 20, 41, 46
Nets 31
Odd and even 7, 29, 44
Operations 12, 16, 17, 22, 30, 36, 40, 44
Rectangles 26
Square numbers 4
Squares 4
Symmetry 34, 35

National Curriculum references for Peak Plus materials

Photocopy

Details of the Attainment Targets, at the various levels, included in the materials for Years 1 and 2, including this book, are shown below.

Suggestions for a rich variety of activities across all the Attainment Targets can be found in *Key Stage 1 First, Second and Third Resource Books*.

Pupil Workcards (Modules 1 to 8)

Attainment Targets	Level 1	Level 2	Level 3
1			
2			
3			
5			
6			
8			
13			

Topic Workbooks

Attainment Targets	Level 1
2	
8	
10	
11	
13	

Number Explorers (1 to 8)

Attainment Targets	Level 2	Level 3
2		
3		
4		
5		
6		

Peak Plus Starters/More Starters

Attainment Targets	Level 2	Level 3
1		
2		
3		
5		
6		
8		
9		
10		
11		
12		
13		
14		

Early Number Games

Attainment Targets	Level 1	Level 2
1/9		